# 木星

## 太阳系的“巨人”

# JUPITER

## The Giant of the Solar System

（英国）埃伦·劳伦斯／著　刘 颖／译

江苏凤凰美术出版社

**著作权合同登记图字：10–2022–144**

图书在版编目（CIP）数据

木星：太阳系的“巨人” /（英）埃伦·劳伦斯著；刘颖译. -- 南京：江苏凤凰美术出版社，2025. 1.
（环游太空）. -- ISBN 978-7-5741-2027-3

Ⅰ. P185.4-49

中国国家版本馆 CIP 数据核字第 2024MT6855 号

| | |
|---|---|
| 策划 | 朱婧 |
| 责任编辑 | 高静　奚鑫 |
| 责任校对 | 王璇 |
| 责任设计编辑 | 樊旭颖 |
| 责任监印 | 生嫄 |
| 英文朗读 | C.A.Scully |
| 项目协助 | 邵楚楚　乔一文雯 |

| | |
|---|---|
| 丛书名 | 环游太空 |
| 书名 | 木星：太阳系的“巨人” |
| 著者 | （英国）埃伦·劳伦斯 |
| 译者 | 刘颖 |
| 出版发行 | 江苏凤凰美术出版社（南京市湖南路 1 号　邮编：210009） |
| 印刷 | 南京新世纪联盟印务有限公司 |
| 开本 | 710 mm × 1000 mm　1/16 |
| 总印张 | 18 |
| 版次 | 2025 年 1 月第 1 版 |
| 印次 | 2025 年 1 月第 1 次印刷 |
| 标准书号 | ISBN 978-7-5741-2027-3 |
| 总定价 | 198.00 元（全 12 册） |

# 目录 Contents

书中加粗的词语见词汇表解释。

Words shown in **bold** in the text are explained in the glossary.

# 欢迎来到木星

# Welcome to Jupiter

想象一下，你正飞去一个距离地球数亿千米的星球。

Imagine flying to a world that is hundreds of millions of kilometers from Earth.

随着太空飞船越飞越近，你将看见旋转的彩色云层。

As your spacecraft gets close, you see colorful swirling clouds.

云层内是呼啸的强风和猛烈的风暴，足以摧毁太空飞船。

Inside the layer of clouds are strong winds and fierce storms that could destroy your spacecraft.

即使你成功穿越了云层，也找不到可着陆的地方。

Even if you make it through the clouds, there is nowhere to land.

因为这个遥远的世界是一个由气体和液体组成的巨大球体。

That's because this faraway world is a gigantic ball of **gases** and liquids.

欢迎来到行星木星！

Welcome to the **planet** Jupiter!

除了太空飞船外，还从未有人去过木星。1973年，一个名为“先驱者10号”的太空探测器成为第一个飞越木星的航天器。

No humans have ever visited Jupiter, but spacecraft have. In 1973, a space **probe** called *Pioneer 10* was the first spacecraft to fly past the planet.

“先驱者10号”拍摄了数百张木星照片，并传输回地球。

*Pioneer 10* took hundreds of photos of Jupiter and beamed them back to Earth.

木星周围的云层
The clouds around Jupiter

# 太阳系 The Solar System

木星以大约47 000千米每小时的速度在太空移动。

Jupiter is moving through space at about 47,000 kilometers per hour.

它围绕着太阳做一个巨大的圆周运动。

It is moving in a huge circle around the Sun.

木星是围绕太阳公转的八大行星之一。

Jupiter is one of eight planets circling the Sun.

八大行星分别是水星、金星、我们的母星地球、火星、木星、土星、天王星和海王星。

The planets are called Mercury, Venus, our home planet Earth, Mars, Jupiter, Saturn, Uranus, and Neptune.

冰冻的彗星和被称为“小行星”的大型岩石也围绕着太阳公转。

Icy **comets** and large rocks, called **asteroids**, are also moving around the Sun.

太阳、行星和其他天体共同组成了“太阳系”。

Together, the Sun, the planets, and other space objects are called the **solar system**.

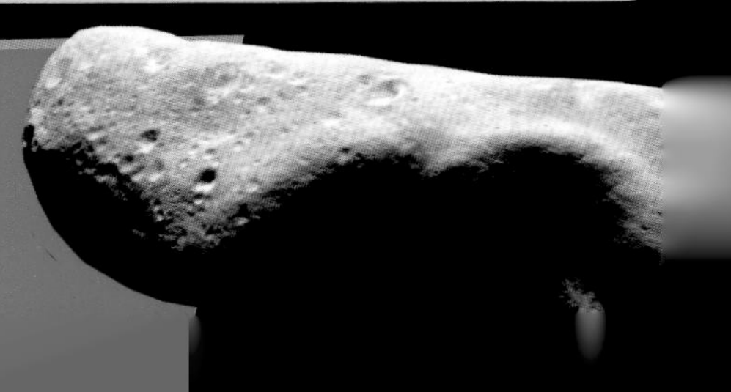

太阳系中的大多数小行星都集中在被称为“小行星带”的环状带中。

Most of the asteroids in the solar system are in a ring called the asteroid belt.

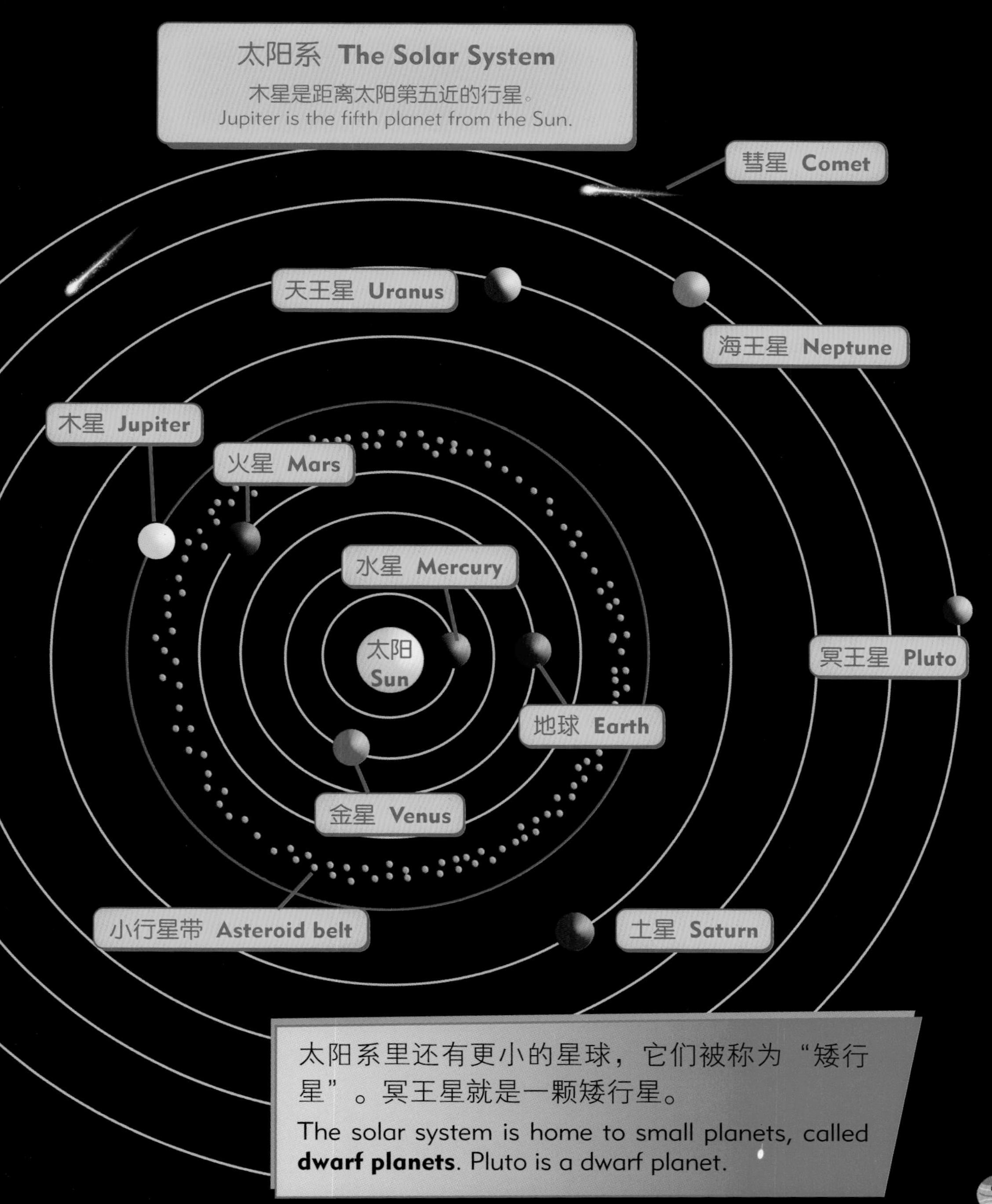


太阳系里还有更小的星球，它们被称为“矮行星”。冥王星就是一颗矮行星。

The solar system is home to small planets, called **dwarf planets**. Pluto is a dwarf planet.

# 木星的奇幻之旅
# Jupiter's Amazing Journey

行星围绕太阳公转一圈所需的时间被称为“一年”。

The time it takes a planet to **orbit**, or circle, the Sun once is called its year.

地球绕太阳公转一圈需要略多于365天的时间，所以地球上的一年有365天。

Earth takes just over 365 days to orbit the Sun, so a year on Earth lasts 365 days.

木星比地球离太阳更远，所以它绕太阳公转的路程长得多。

Jupiter is farther from the Sun than Earth, so it must make a much longer journey.

木星绕太阳公转一圈大约需要12个地球年。

It takes Jupiter nearly 12 Earth years to orbit the Sun.

这意味着，地球上的12年才相当于木星上的1年！

This means that a 12-year-old on Earth would just be turning 1 in Jupiter years!

当行星围绕太阳公转时，它也像陀螺一样自转着。

As a planet orbits the Sun, it also spins, or **rotates**, like a top.

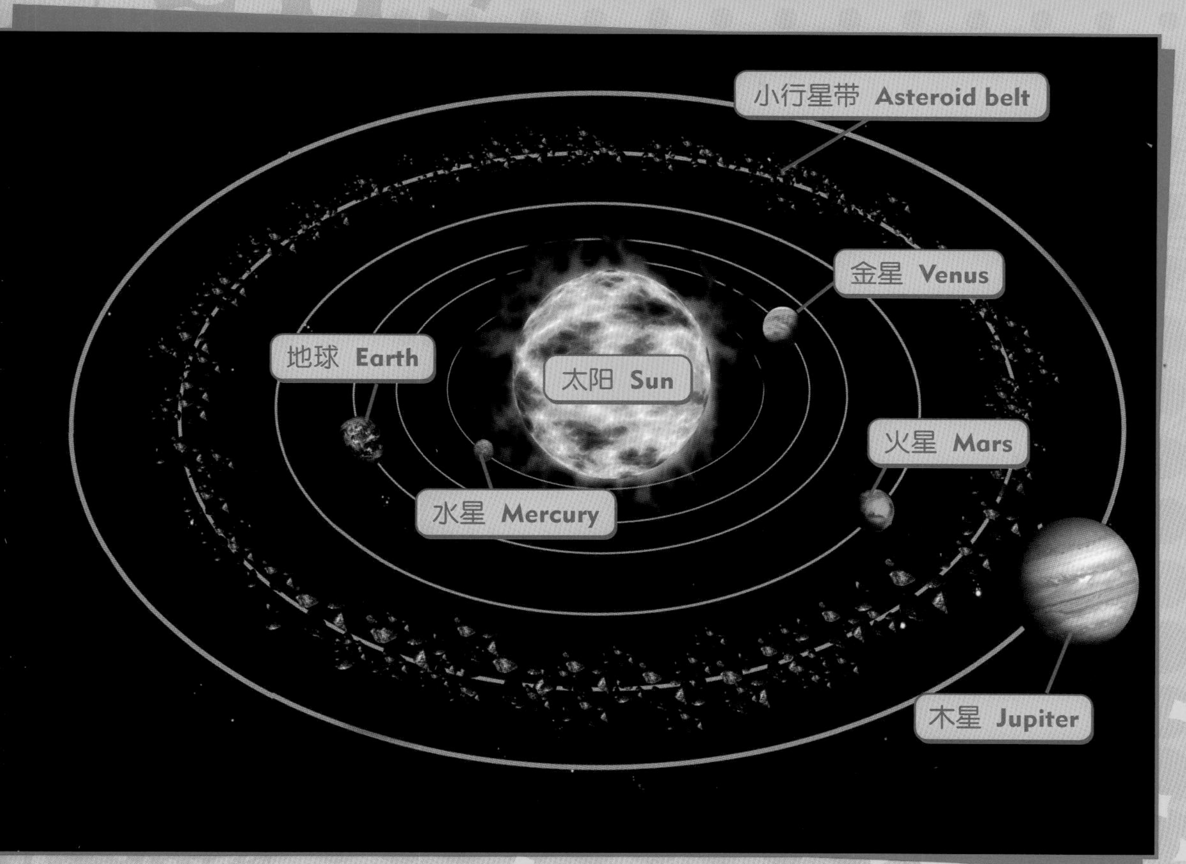


地球绕太阳公转一圈的路程约为9.4亿千米，而木星绕太阳公转一圈的距离约为50亿千米。

To orbit the Sun once, Earth makes a journey of about 940 million kilometers. Jupiter must make a journey of about 5 billion kilometers.

# 近距离观察木星

# A Closer Look at Jupiter

木星是太阳系中迄今最大的行星。

Jupiter is by far the largest planet in the solar system.

它的直径是地球的11倍！

It is 11 times as wide as Earth!

不同于岩石构成的地球，木星没有固体表面。

Unlike Earth, which is a rocky planet, Jupiter doesn't have a solid surface.

木星的大气层是一层厚厚的气体，外层是云层。

Jupiter has an **atmosphere** that is a thick layer of gases with an outer layer of clouds.

在大气层之下，这颗行星是由液体组成的巨大球体。

Beneath its atmosphere, the planet is a giant ball of liquids.

## 木星是由什么构成的？
## What Is Jupiter Made Of?

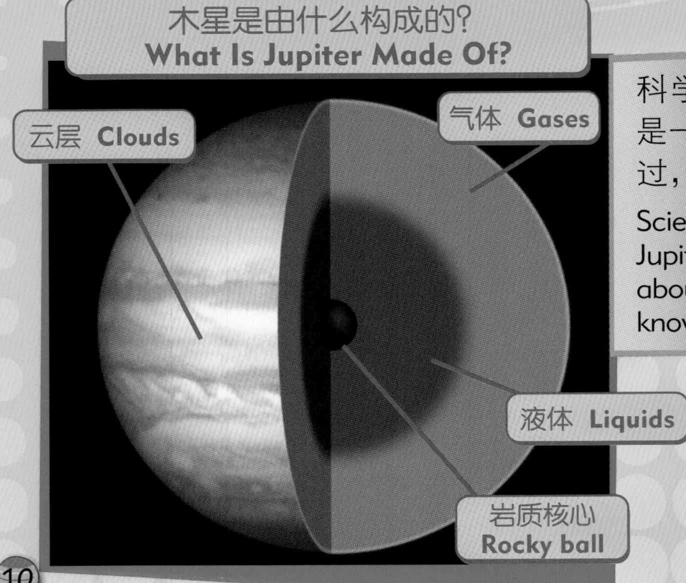


科学家认为，木星的中心可能是一个地球大小的岩石核心。不过，这个猜想尚未得到证明。

Scientists think that in the center of Jupiter there may be a ball of rock about the size of Earth. No one knows for sure, though.

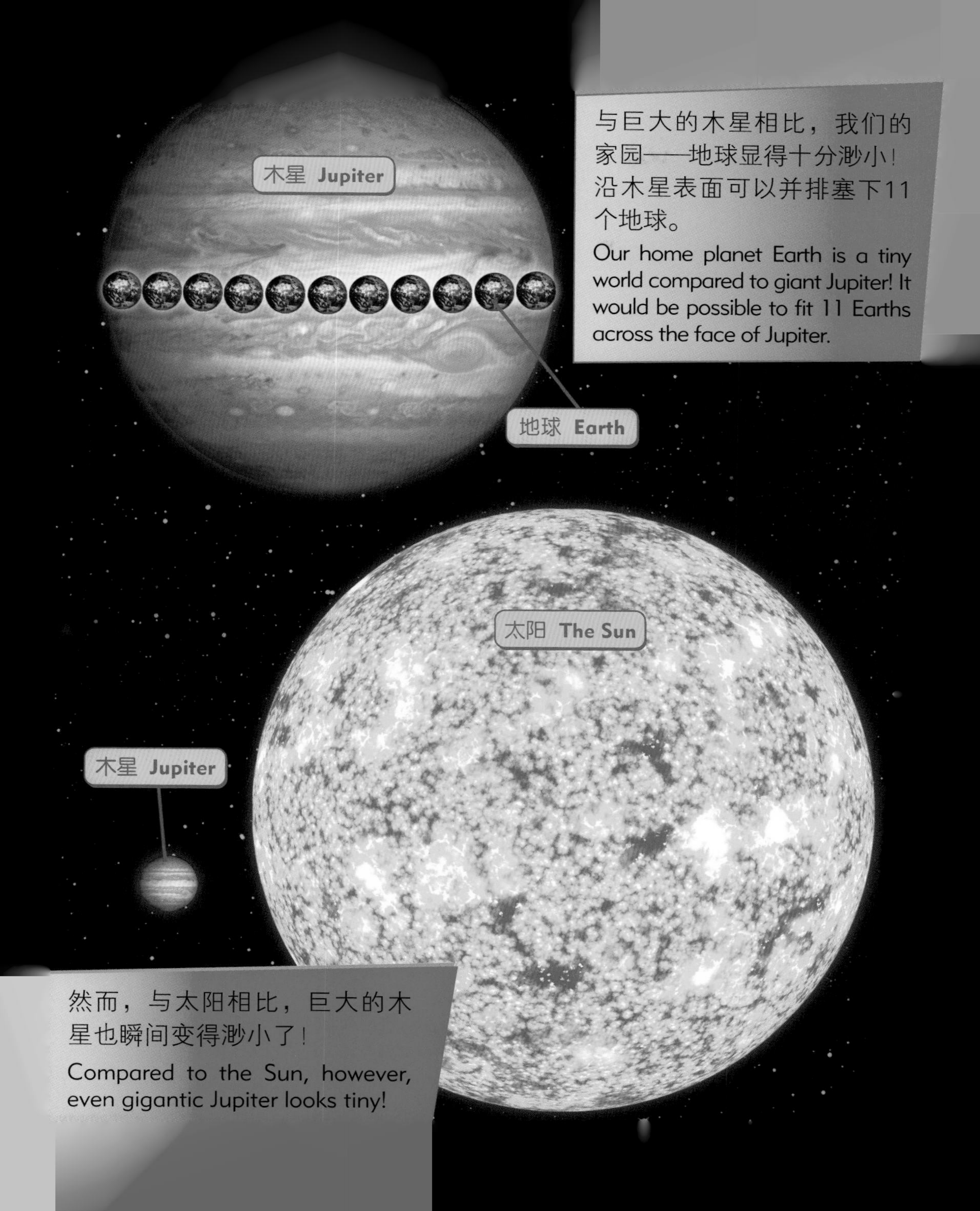
与巨大的木星相比，我们的家园——地球显得十分渺小！沿木星表面可以并排塞下11个地球。
Our home planet Earth is a tiny world compared to giant Jupiter! It would be possible to fit 11 Earths across the face of Jupiter.
木星 Jupiter
地球 Earth
太阳 The Sun
木星 Jupiter
然而，与太阳相比，巨大的木星也瞬间变得渺小了！
Compared to the Sun, however, even gigantic Jupiter looks tiny!

# 大红斑
# The Great Red Spot

木星上有片巨大的风暴，被称为“大红斑”。

Jupiter is home to a huge storm called the Great Red Spot.

大红斑就像巨大的旋转飓风。

The Great Red Spot is like an enormous, spinning **hurricane**.

大到在地球上使用望远镜就能清楚地看见它。

It is so big that it can easily be seen from Earth through a telescope.

事实上，人们第一次看见大红斑是在400年前发明望远镜的时候。

In fact, people first saw the Great Red Spot when telescopes were invented 400 years ago.

这意味着，这片超级飓风持续了至少400年！

This means the super-size hurricane has been going on for at least 400 years!

这张图显示了大红斑与地球的大小对比。

This picture shows the size of the Great Red Spot compared to Earth.

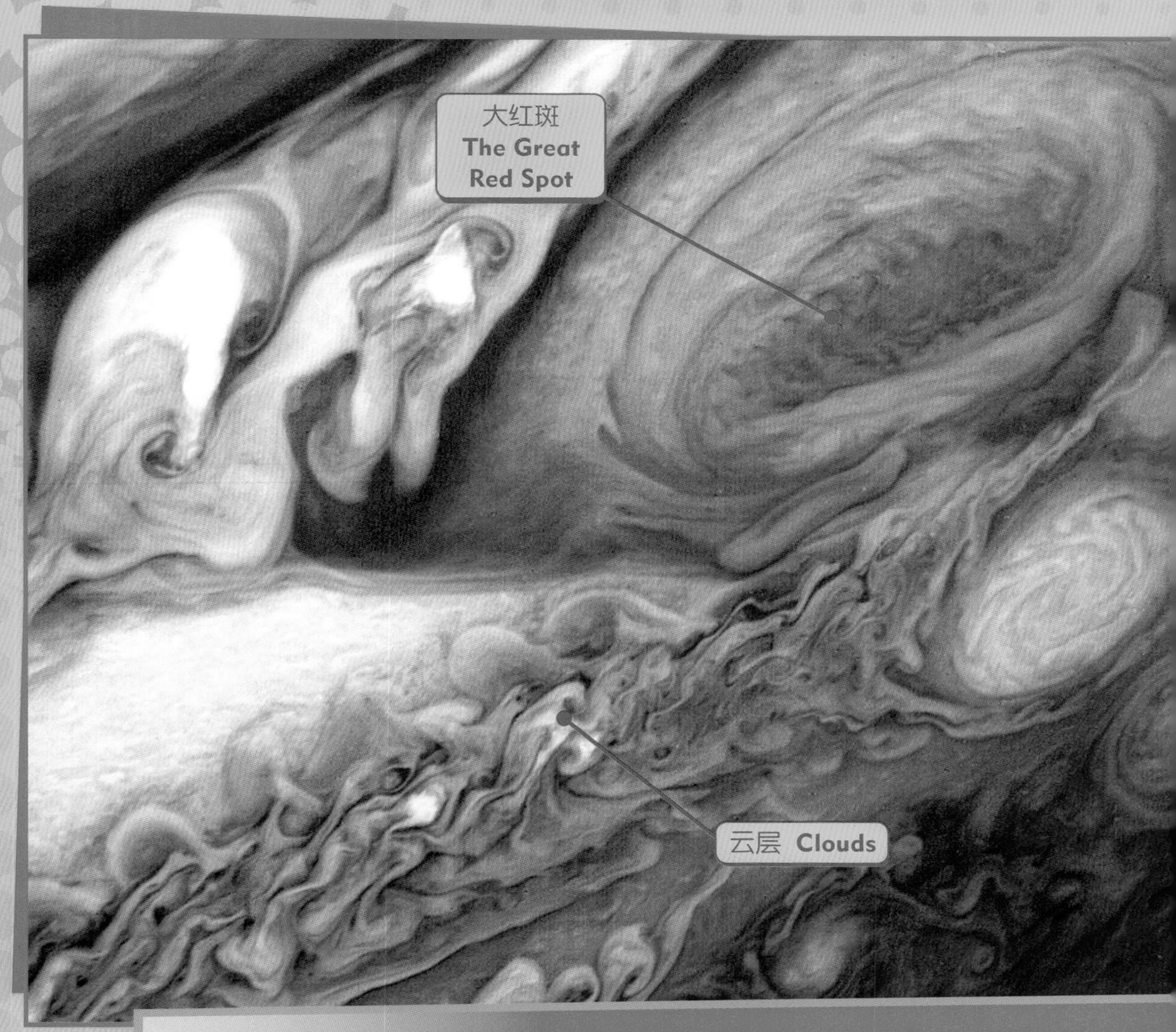


这张照片是太空探测器“旅行者1号”拍摄的。照片里是木星的云层和大红斑。云层中五颜六色的旋涡图案是由气体组成的。

This picture was taken by a space probe called *Voyager 1*. It shows Jupiter's clouds and the Great Red Spot. The swirling, colorful patterns in the clouds are made by gases.

# 木星家族
# Jupiter's Family

木星和它周围的小星球共同组成了一个大家庭。

Jupiter has an amazing family of smaller worlds circling it.

这些由岩石和冰块构成的天体是这颗巨行星的卫星。

These rocky and icy space objects are the giant planet's **moons**.

我们的家园——地球，只有1颗卫星。

Earth, our home planet, has just one moon.

木星至少有95颗卫星，而且科学家还在发现新的卫星。

Jupiter has at least 95 moons, and scientists are still discovering new ones!

环绕木星的还有4圈由尘埃和小块岩石构成的行星环。

Also circling Jupiter are four rings of dust and small pieces of rock.

## 木星最大的卫星
## Jupiter's Largest Moon

木星最大的卫星是盖尼米得（木卫三）。它是整个太阳系中最大的卫星。它甚至比水星还大。

Jupiter's largest moon is called Ganymede. It is the biggest moon in the whole solar system. It is even bigger than the planet Mercury.

盖尼米得被一层厚厚的冰覆盖，冰层厚度约800千米！

Ganymede is covered with a layer of ice that may be 800 kilometers thick!

## 木星的四大卫星
## Jupiter's Four Largest Moons

盖尼米得（木卫三）
**Ganymede**
**(GA-neh-mede)**

卡里斯托（木卫四）
**Callisto**
**(kuh-LIS-toh)**

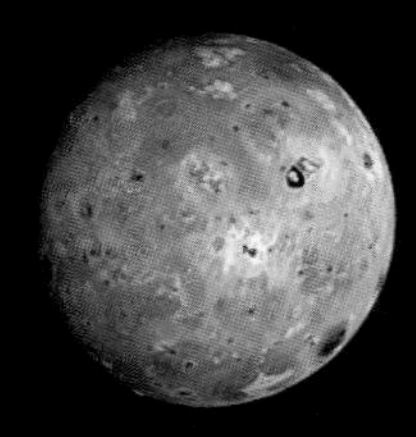

艾奥（木卫一）
**Io**
**(EYE-oh)**

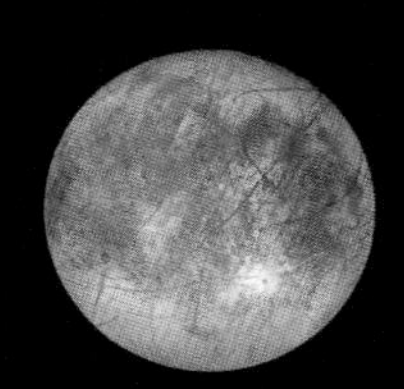

欧罗巴（木卫二）
**Europa**
**(yoo-ROH-puh)**

# 神奇的卫星
# Amazing Moons

飞往木星的太空飞船还揭示了许多有关木星卫星的信息。

The spacecraft that visited Jupiter also discovered lots of information about the planet's moons.

木星的卫星之一——卡里斯托，它的地面上布满了陨石坑。

The ground on one moon, Callisto, is covered with large holes called craters.

木星的另一颗卫星——艾奥（木卫一），它有超过400座火山。

Jupiter's moon Io is home to more than 400 **volcanoes**.

欧罗巴（木卫二）的表面是一片深海。

The whole surface of Europa is covered with a deep ocean.

然而，欧罗巴非常寒冷，以至于海面都结冰了！

It is so cold on Europa, though, that the surface of the ocean is frozen!

卡里斯托上的陨石坑是由小行星等岩石天体撞击而形成的。

The craters on Callisto were made by rocky objects, such as asteroids, that hit the moon.

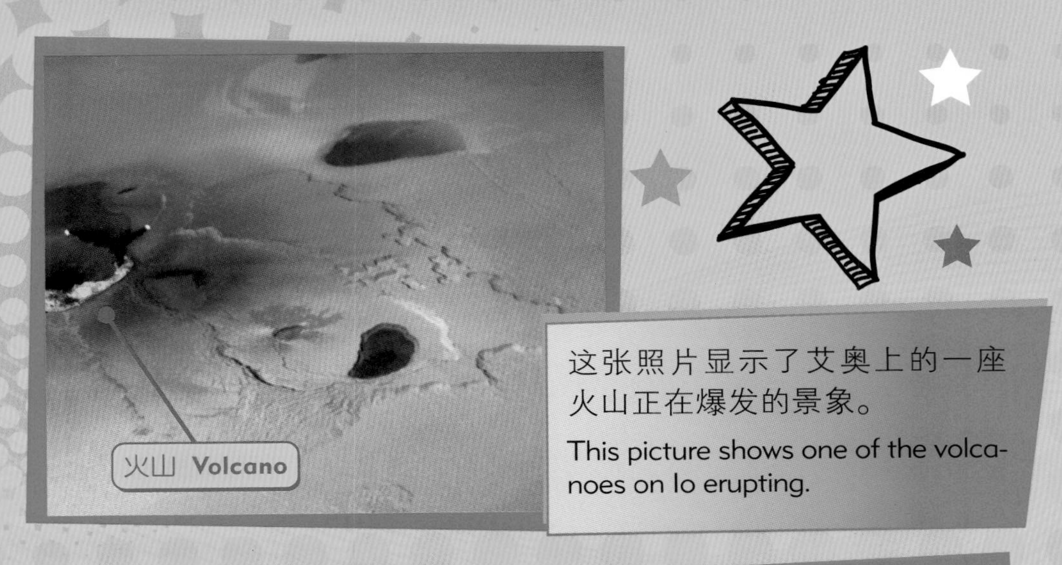


这张照片显示了艾奥上的一座火山正在爆发的景象。

This picture shows one of the volcanoes on Io erupting.

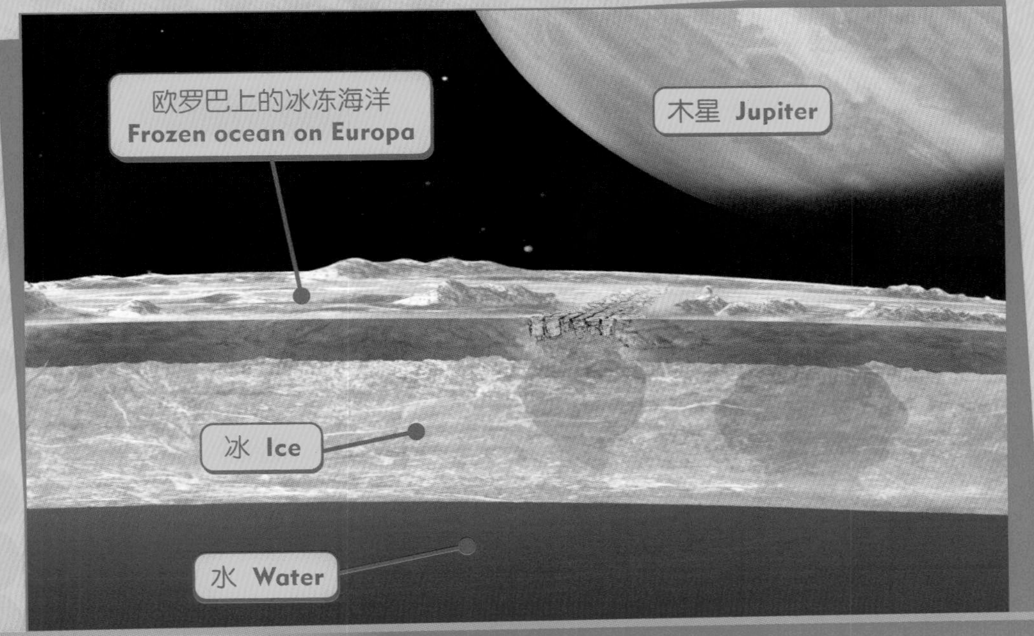

这张图显示了欧罗巴表面的样子。科学家认为，在冰封的海面下可能有水。

This picture shows how Europa's surface might look. Scientists think that under the top layer of frozen ocean, there could be water.

# 探测木星的任务

## Missions to Jupiter

1989年，太空探测器“伽利略号”离开地球，前往木星执行探测任务。

In 1989, a space probe named *Galileo* (gal-uh-LAY-oh) left Earth on a mission to Jupiter.

“伽利略号”围绕木星飞行了近8年，研究木星及其卫星。

*Galileo* orbited Jupiter for nearly eight years, studying the planet and its moons.

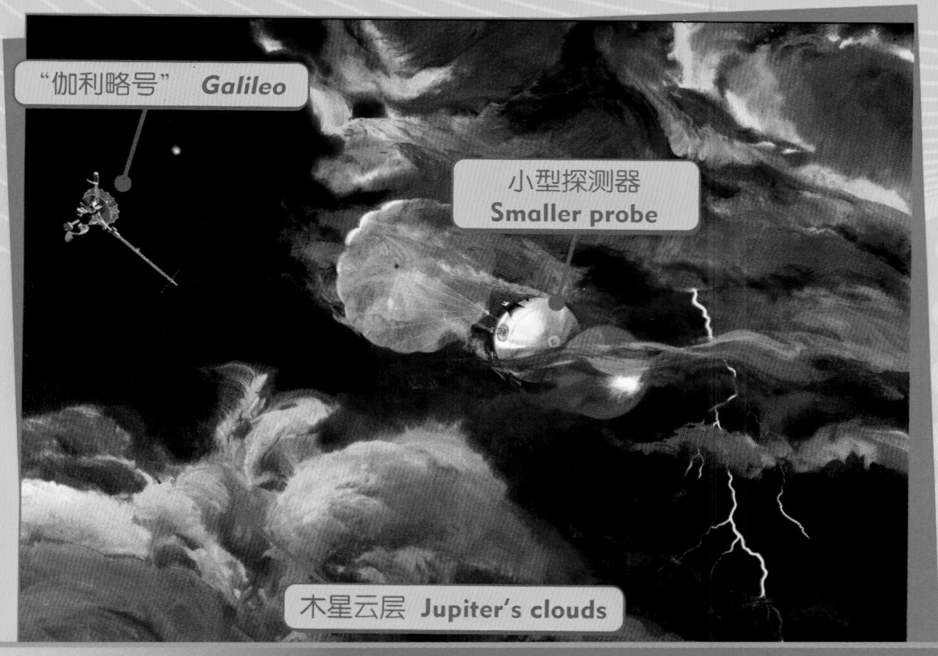


这张图显示了“伽利略号”向木星的云层和大气层发射了一个小型探测器。探测器将信息传输回“伽利略号”，这一过程持续了58分钟。

This picture shows *Galileo* sending a smaller probe into Jupiter's clouds and atmosphere. The probe beamed information back to *Galileo* for 58 minutes.

2016年，探测器“朱诺号”抵达木星。

In 2016, a probe named *Juno* arrived at Jupiter.

它让科学家观测到了木星云层之下的景象。

It has let scientists see beneath Jupiter's clouds.

“朱诺号”研究了木星的风、大气，以及这颗星球的内核是否有岩石存在。

*Juno* has studied Jupiter's winds, its atmosphere, and if it has rock in its center.

2021年，“朱诺号”开始了预计3年的探测任务，去造访盖尼米得、欧罗巴、艾奥等木星卫星。它会飞行在这些卫星上空仅几百千米的位置。

In 2021, *Juno* began a three year mission to visit Ganymede, Europa, and Io. It flew just a few hundred kilometers above the surface of the moons.

2019年的木星
**Jupiter in 2019**

本页的所有照片都是由“朱诺号”拍摄的。

The images on this page were taken by *Juno*.

此照片展示了木星南极的风暴。

This image shows storms at Jupiter's South Pole.

# 有趣的木星知识
# Jupiter Fact File

以下是一些有趣的木星知识：木星是距离太阳第五近的行星。

Here are some key facts about Jupiter, the fifth planet from the Sun.

## 木星的发现
## Discovery of Jupiter

不用望远镜也能在天空中看见木星。人们早在古代就发现了木星。

Jupiter can be seen in the sky without a telescope. People have known it was there since ancient times.

## 木星是如何得名的
## How Jupiter got its name

木星的命名是由古罗马众神之王得来的。

The planet is named after the king of the Roman gods.

## 行星的大小
## Planet sizes

这张图显示了太阳系八大行星的大小对比。

This picture shows the sizes of the solar system's planets compared to each other.



## 木星的大小
## Jupiter's size

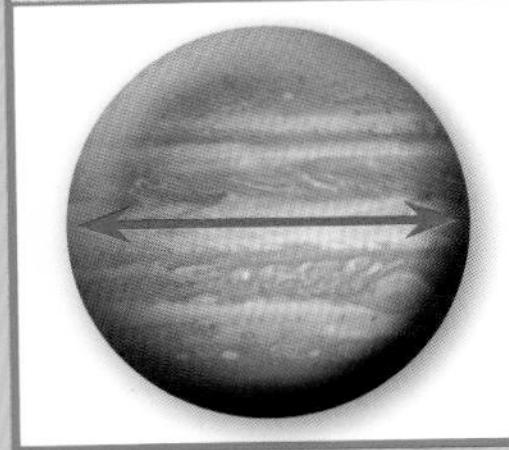

木星的直径约139 822千米

About 139,822 km across

## 木星自转一圈需要多长时间
## How long it takes for Jupiter to rotate once

将近10个地球时

Nearly 10 Earth hours

## 木星与太阳的距离
## Jupiter's distance from the Sun

木星与太阳的最短距离是740 679 835千米。

木星与太阳的最远距离是816 001 807千米。

The closest Jupiter gets to the Sun is 740,679,835 km.

The farthest Jupiter gets from the Sun is 816,001,807 km.

## 木星绕太阳的轨道长度
## Length of Jupiter's orbit around the Sun

4 887 595 931千米

4,887,595,931 km

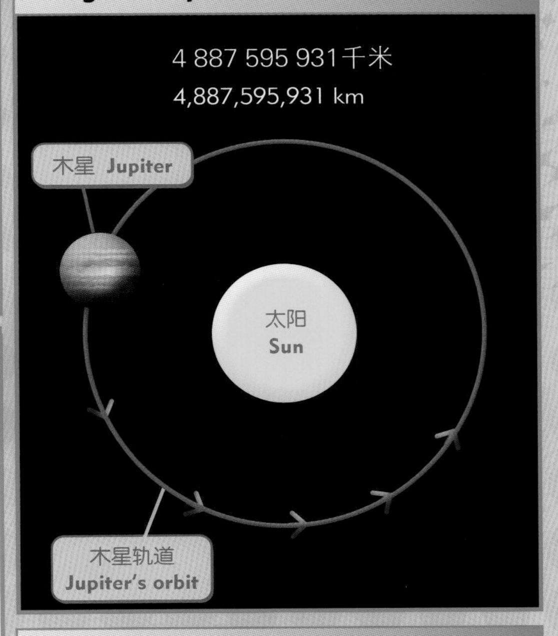


## 木星围绕太阳公转的平均速度
## Average speed at which Jupiter orbits the Sun

每小时47 002千米

47,002 km/h

## 木星上的一年
## Length of a year on Jupiter

将近4 333个地球天
（约12个地球年）

Nearly 4,333 Earth days
(nearly 12 Earth years)

## 木星的卫星
## Jupiter's moons

木星至少有95颗卫星。还有更多卫星有待发现。

Jupiter has at least 95 moons. There are possibly more waiting to be discovered.

## 木星上的温度
## Temperature on Jupiter

零下148摄氏度

-148°C

# 动动手吧：熔化的蜡笔木星图

## Get Crafty : Melted Wax Crayon Jupiter

使用熔化的蜡笔，在纸上绘制出木星及其流动的彩色云层。

**你需要：**

- 蜡纸
- 剪刀
- 黄色、棕色和橙色蜡笔
- 奶酪刨
- 熨斗（在成年人的帮助下使用）

1. 从蜡纸上剪下两个餐盘大小的圆。

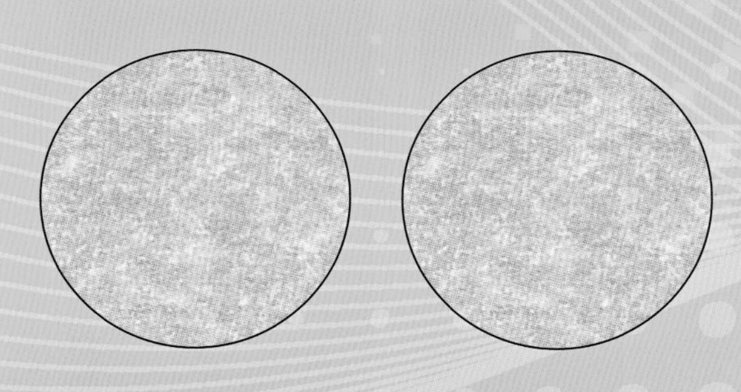

2. 使用奶酪刨将蜡笔屑刨在其中一张圆形蜡纸上。（操作时应有大人监督，务必注意手指安全！）

3. 将另一张蜡纸盖上去，做成蜡笔“三明治”。

4. 请爸爸妈妈帮忙熨烫蜡笔“三明治”，直至蜡笔屑熔化并混合在一起。

5. 将熔化的蜡笔木星图挂在窗前，沐浴在阳光下，或者用它制作一幅星空图。

# 词汇表 Glossary

**小行星 | asteroid**

围绕太阳公转的大块岩石，有些小得像辆汽车，有些大得像座山。

**大气层 | atmosphere**

行星、卫星或恒星周围的一层气体。

**彗星 | comet**

由冰、岩石和尘埃组成的天体，围绕太阳公转。

**矮行星 | dwarf planet**

围绕太阳运行的圆形或近圆形天体，比八大行星小得多。

**气体 | gas**

无固定形状或大小的物质，如氧气或氦气。

**飓风 | hurricane**

巨大的风暴，环绕风暴中心（风眼）旋转。直径可达数百千米，风速可高达每小时320千米。

**卫星 | moon**

围绕行星运行的天体。通常由岩石或岩石和冰构成。直径从几千米到几百千米不等。地球有一个卫星，名为“月球”。

## 公转 | orbit

围绕另一个天体运行。

## 行星 | planet

围绕太阳公转的大型天体：一些行星，如地球，主要是由岩石组成的；其他的行星，如木星，主要是由气体和液体组成的。

## 探测器 | probe

不载人太空飞船。通常被送往行星或其他天体，用于拍摄照片并收集信息，由地球上的科学家操作控制。

## 自转 | rotate

物体自行旋转的运动。

## 太阳系 | solar system

太阳和围绕太阳公转的所有天体，如行星及其卫星、小行星和彗星。

## 火山 | volcano

地下岩浆喷出地表形成的山丘，部分火山会有高温的液态岩石和气体从开口处喷发；存在于行星或其他天体上。